HOTEL
新空间——酒店

HOTEL
新空间——酒店

《新空间》编辑组 编
张晨 译

辽宁科学技术出版社

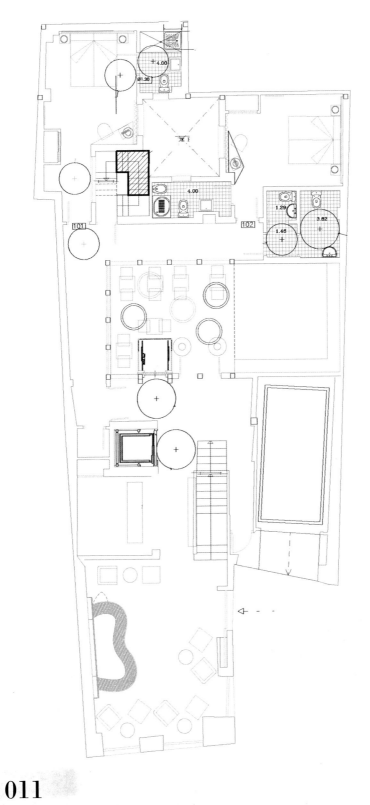

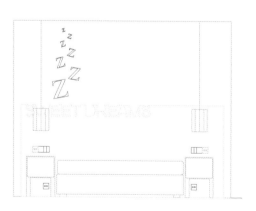
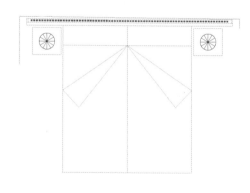

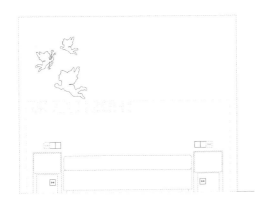 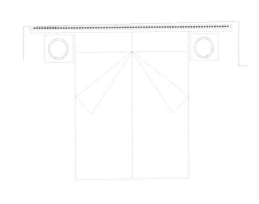

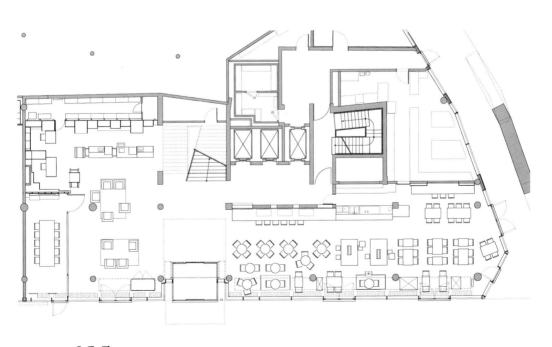

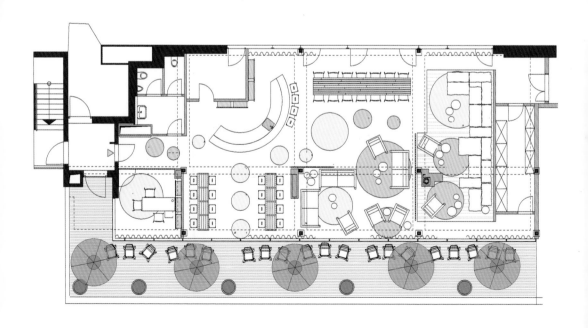

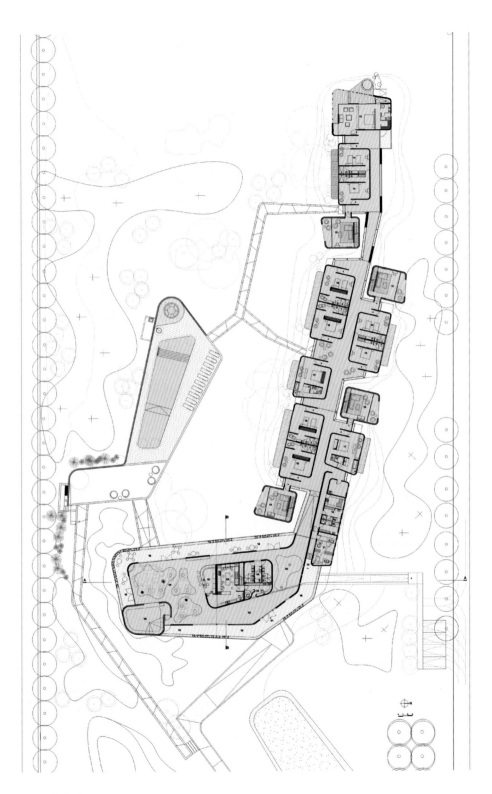

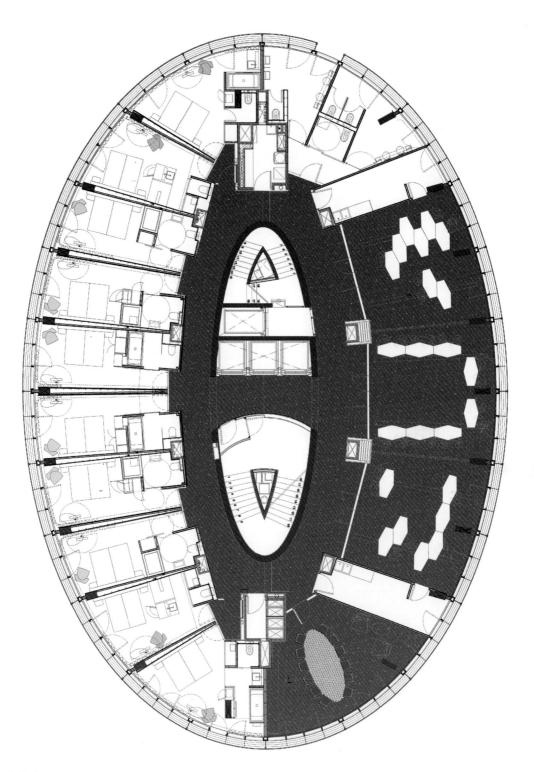

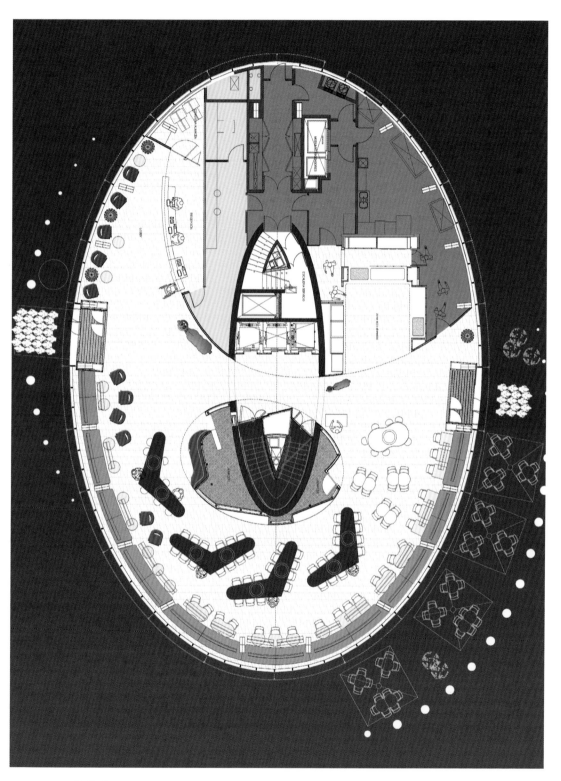

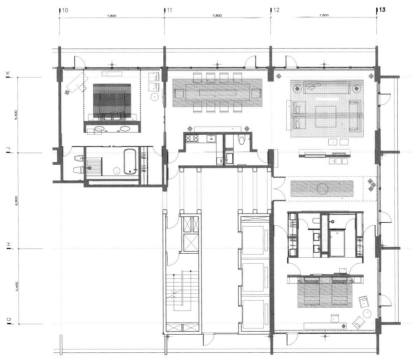

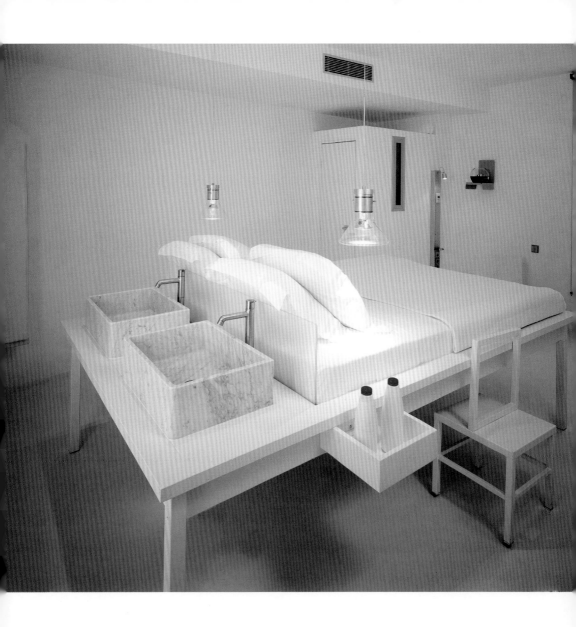

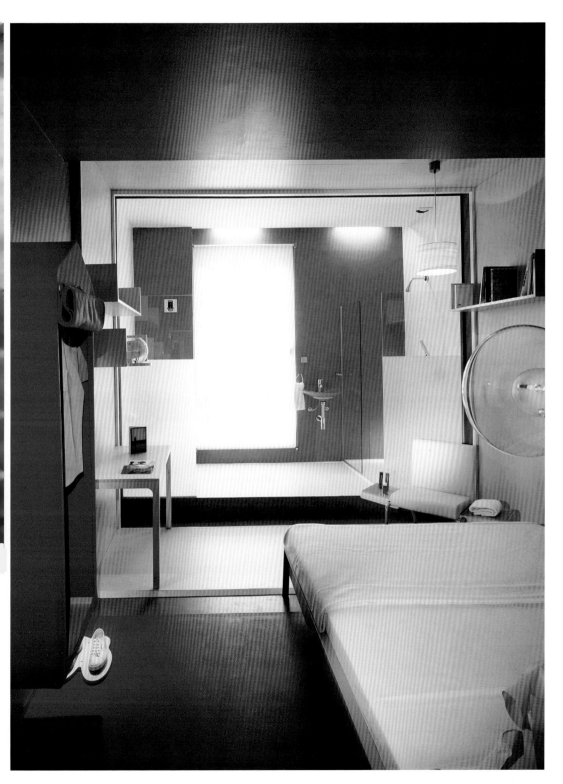

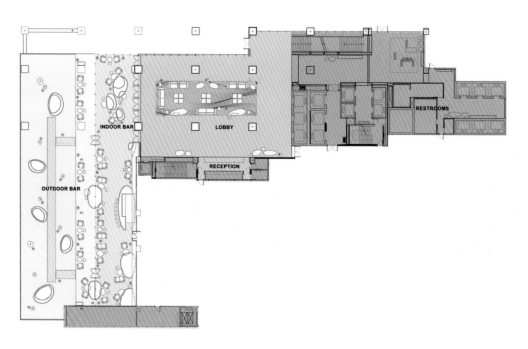

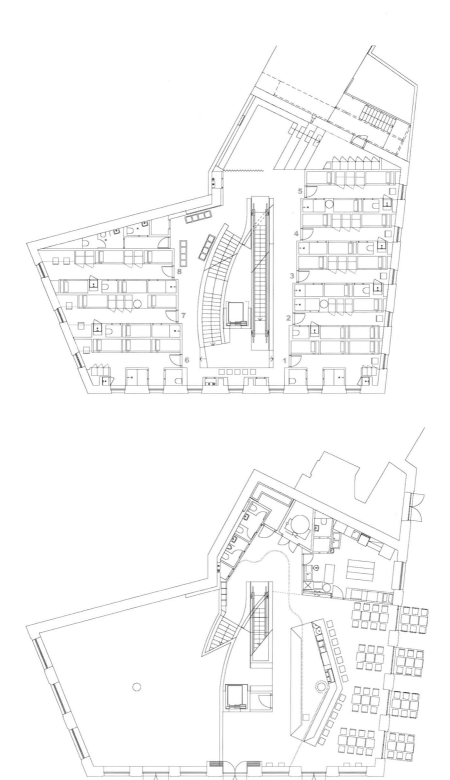

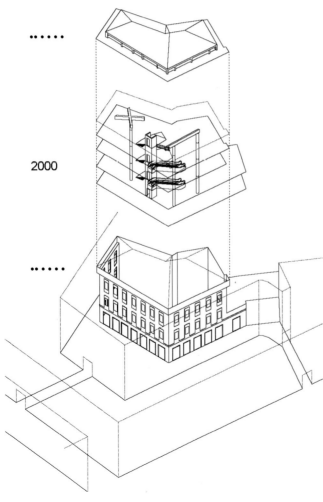
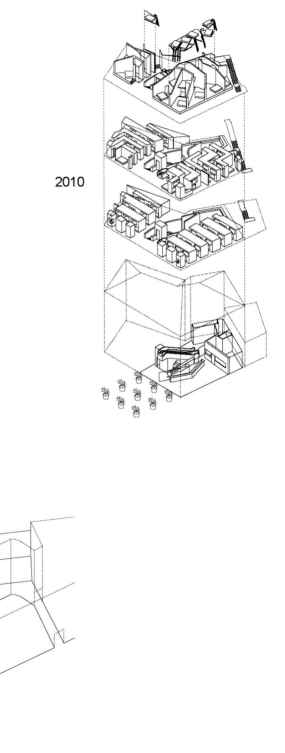

2000

2010

hotel

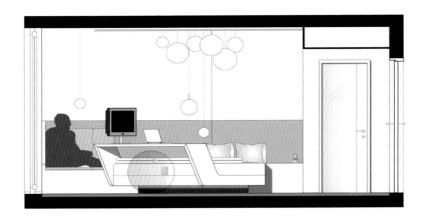

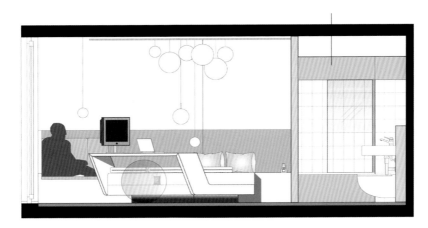

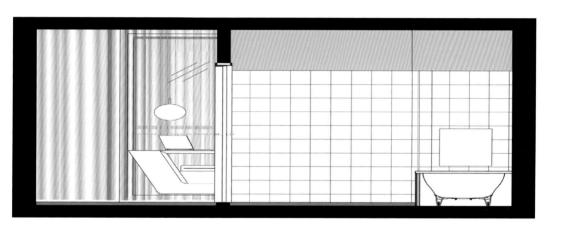
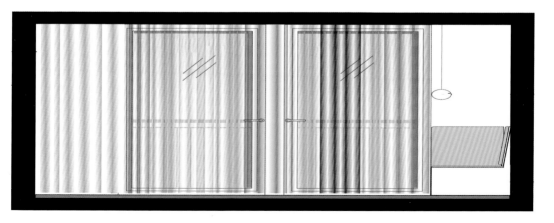

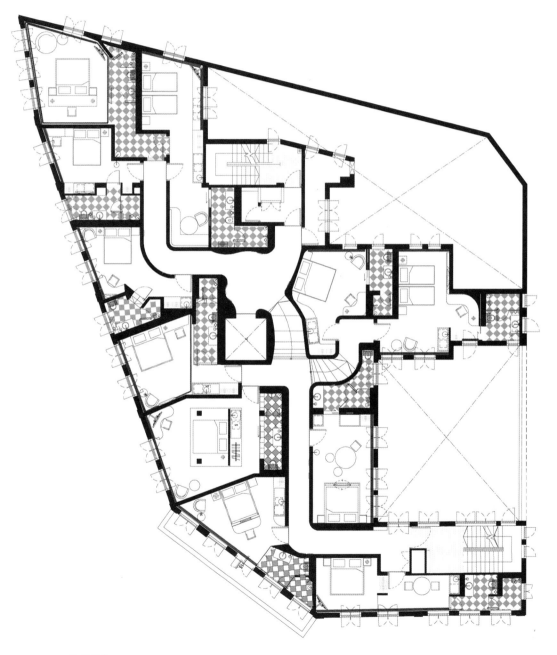

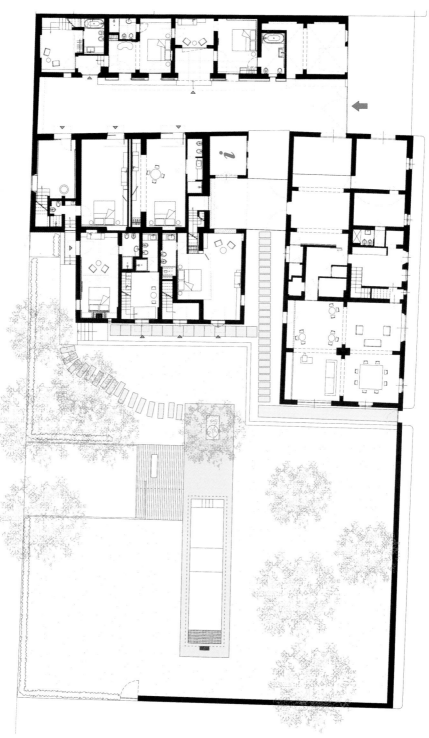

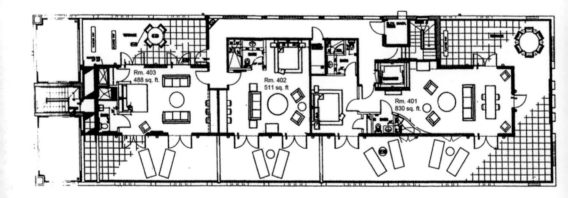

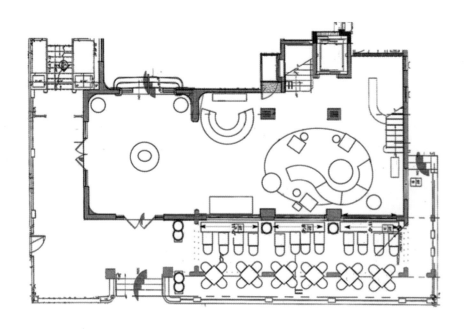

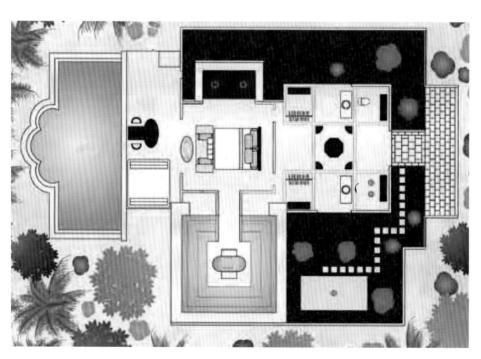

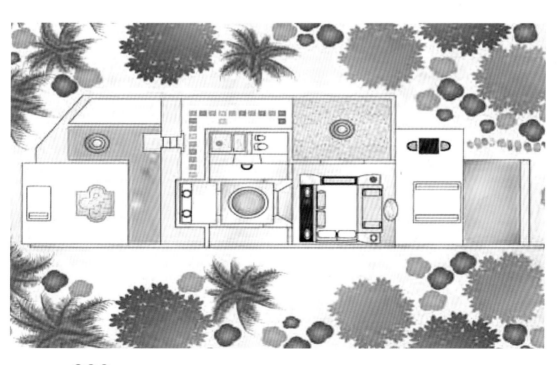

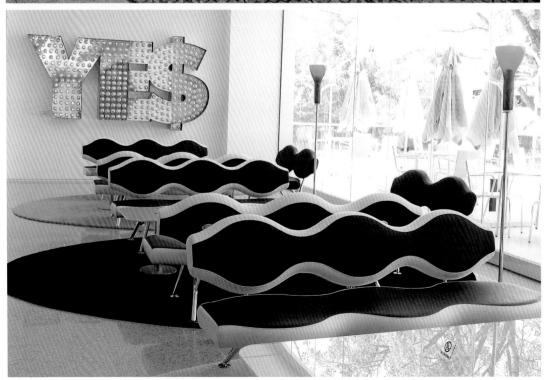

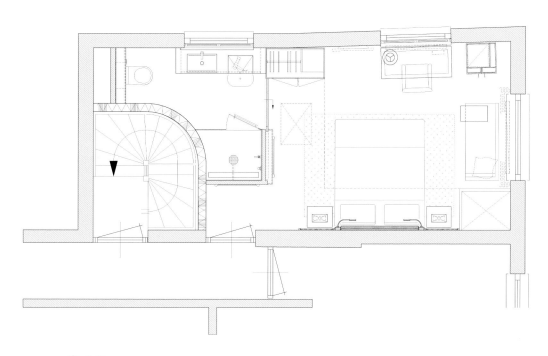

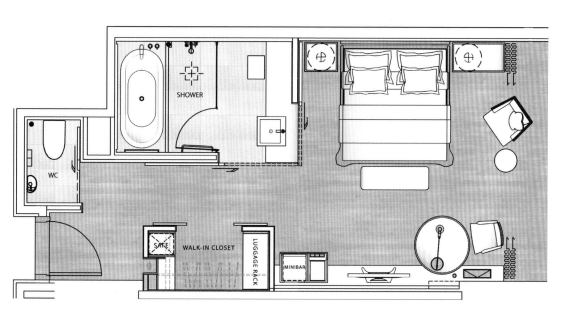

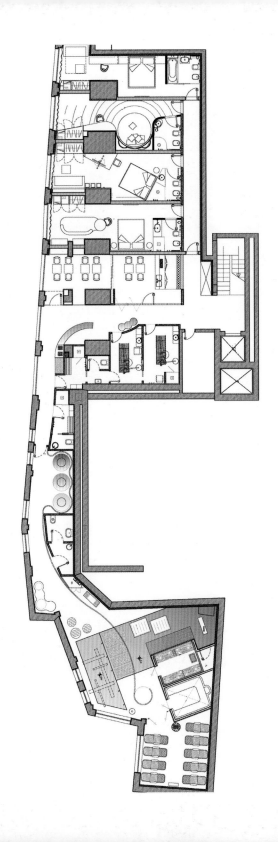

hotel

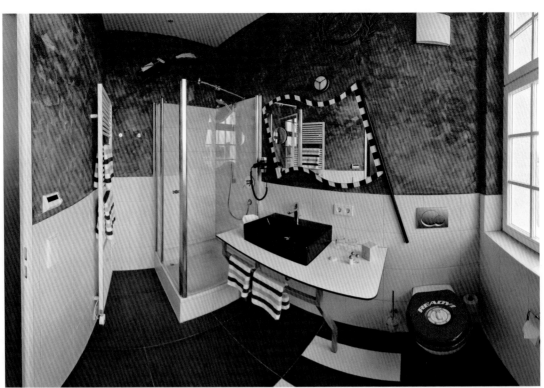

Index
索引

b=bottom, t=top
Designed by

A4 estudio, 88-92

AKSL arhitekti, 48-51

Alberto Apostoli, 108(b), 246-247, 256-257, 274-277

Alessandro Mendini, 52-55

Ander Aginako & Amaia Celaya (abar architects), 150-153

Armin Fischer / dreimeta, 20-29

ASC•Interior, 42-43

Atelier d'architecture King Kong, 169 (b), 170

Atelier Nini Andrade Silva, 236-238

BHDM, 206-213

Brennan Beer Gorman Architects & Chulasai, 61

Cachet Design Studio, 204-205

Camp, 30-33

Carlo Rampazzi, Selvaggio, 76-79

Cheryl Rowley Design, Inc., 108 (t), 109-111

Christian Lacroix, 164-166, 188-189

Christina Zeidler an artist and filmmaker, 102-107

Datumzero Design Office, 168

DBI Design, 258-259

Department of ARCHITECTURE Co., Ltd., 132-137

DIN interiorismo - Aurelio Vázquez, 190-193

dreimeta – Armin Fischer, 252-255

dwp | design worldwide partnership, 214-215

FG stijl, 112-116, 118-119, 260-263

GA Design International (UK), 58, 62(l), 63

Gabellini Sheppard Associates, 82-87

Graven Images, 282-283

Guiliana Salmaso, 120-125

Handel Architects LLP, Frank Fusaro, AIA (Partner), 72-75

Handel Architects LLP / Glenn Rescalvo, AIA (Partner), 292-293

ILMIODESIGN, 4-13

Isay Weinfeld, 248-249

Jestico + Whiles, 182-185

JOI-Design, 250-251

Jordi Galí & Estudi, 93-101

Karim Rashid Inc., 226-229, 242-245

Knevel Architecten, 296-299

lagranja design studio, 180-181

LTW Designworks Pte Ltd., 194-199

Magnus Månsson, 40-41

Matali Crasset, 126-131

Miaja Design Group, 60, 62(t)

Michel Penneman Interior Design, 167, 169(t), 171

Ministry of Design, 172-179

Mr. Duangrit Bunnag / Duangrit Bunnag Architect Limited – DBALP, 264-269

Mr.Sanjay Puri & Mr.Nimish Shah, 44-47

Ohannes K. Wortmann / Wortmann Architects, 162-163

P Landscape, 230-233

PIA, 270-273

Roman-vrtiska, 216-221

Salvatore Ferragamo, Michele Bönan, Nino Solazzi, 234-235

SB Architects, 36-39

Seyhan Ozdemir & Sefer Caglar, 64-65

Söhne & Partner architects, 239-241

SPACE Architects & Designers, 278-281

Stella Cadente, 200-203

Stephen Williams Associates, 14-19

Studio 63 Architecture + Design, 80-81, 117

Studio lot, Achim M. Kammerer, 154-161

Studio Marco Piva, Studio Bam Design, 222-225

STUDIO UP, 140-149

The Buchan Group, 138-139

Tonino Cacace, 56-57, 59

Urban Studio, 66-71

V8 Hotel in Meilenwerk, 284-291

W Global Brand Design & Rockwell Group Europe (RGe), 294-295

Zoltán Tima I Tima Studio - Kozti Zrt, 186-187

Index
索引

Photographs have been obtained from the designers. Thanks photographers are listed below.
b=bottom, t=top

Alfonso Acedo, 4-13

Ali Bekman, 64-65

Antje Hanebeck, 154-161

Arthur Pequin, 169 (b), 170

Arturo Chávez, 190-193

Bruce Damonte, Phillip Ennis, 72-75

Byblos Art Hotel Villa Amista, 52-55

Capri Palace Hotel & Spa, 56-57, 59

Centara Grand Island Resort & Spa Maldives, 60, 62(t)

Centara Hotels and Resorts Photographer, 61

Cheryl Rowley Design, Inc, 108 (t), 109-111

Christopher Cypert, 230-233

Christopher Frederick Jones and Hilton Hotels & Resorts, 138-139

CI&A Photography, 172-179

Ciro Ceolho, 36-39

Courtesy of Armin Fischer / dreimeta, 20-29

Courtesy of camp, 30-33

Courtesy of Carlo Rampazzi, Selvaggio, 76-79

Courtesy of Christian Lacroix, 164-166

Courtesy of Christian Lacroix, 188-189

Courtesy of FG stijl, 260-263

Courtesy of Isay Weinfeld, 248-249

Courtesy of Karim Rashid Inc., 226-229, 242-245

Courtesy of Lords South Beach, 206-213

Courtesy of Matali Crasset, 126-131

Courtesy of Michel Penneman Interior Design, 167, 169(t), 171

Courtesy of Stella Cadente, 200-203

Courtesy of Urban Studio, 66-71

Duangrit Bunnag Architect Limited – DBALP, 264-269

dwp | design worldwide partnership, 214-215

Filip Slapal, 216-221

Frank Hoppe, 284-291

GARCIA&BETANCOURT, 88-92

Gary Graves + Anne Gridley, 168

Gerry O'leary, 182-185

Gladstone Hotel, 102-107

Hans Fonk, James Stokes, 112-116, 118-119

Henrique Seruca, Alma Mollemans, 236-238

Ian Gibb, 270-273

JOI-Design, 250-251

Jordi Miralles, 93-101

José Manuel Cutillas, Pako Vierbücher, 150-153

Krister Engström Grafia, 40-41

lagranja design studio, 180-181

Luca Morandini, 108(I), 246-247, 256-257, 274-277

Lungarno Collection, 234-235

Luuk Kramer, 296-299

Marc Gerritsen, 194-199

Mark Seelen, 222-225

Minchaya Chayosumrit, Best Western Plus @20 Sukhumvit Hotel, 42-43

Miran Kambič, 48-51

MİRHAN BİLİR, 278-281

Richard Powers, 58, 62(b), 63

Robert Leš, 140-149

Severin Wurnig, 239-241

Silvestrin/Salmaso, 120-125

Spaceshift Studio, 204-205

Starwood, 294-295

Stefan Müller, 162-163

Stephen Williams Associates, 14-19

Steve Herud, 252-255

Tamás Bujnovszky, 186-187

Teviot Creative, 282-283

The Istanbul Edition Hotel, 82-87

Tony Phillips, Brent Winstone & Tyrone Brannigan, 258-259

Vinesh Gandhi, 44-47

W Santiago, 292-293

Wison Tungthunya, 132-137

Yael Pincus, 80-81, 117

图书在版编目（CIP）数据

新空间. 酒店 / 新空间编辑组编；张晨译.
- 沈阳：辽宁科学技术出版社，2016.3
ISBN 978-7-5381-9551-4

Ⅰ.①新… Ⅱ.①新… ②张… Ⅲ.①饭店－室内装饰设计－世界－图集 Ⅳ.①TU238-64

中国版本图书馆 CIP 数据核字(2016)第 013607 号

出版发行：辽宁科学技术出版社
　　　　（地址：沈阳市和平区十一纬路29号 邮编：110003）
印　刷　者：利丰雅高印刷（深圳）有限公司
经　销　者：各地新华书店
幅面尺寸：170mm×225mm
印　　　张：19
插　　　页：4
出版时间：2016年 3 月第 1 版
印刷时间：2016年 3 月第 1 次印刷
责任编辑：殷　倩
封面设计：周　洁
版式设计：周　洁
责任校对：周　文

书　　号：ISBN 978-7-5381-9551-4
定　　价：88.00元

联系电话：024-23284360
邮购热线：024-23284502
http://www.lnkj.com.cn